An Algebra Primer: What You Need to Know *Before* You Can Solve for X

MARK PHILLIPS

Published by A. J. Cornell Publications

ISBN: 978-0-9850501-4-6

CONTENTS

1 TERMS AND SYMBOLS YOU NEED TO KNOW

Learning a new type of mathematics is a little like learning a new language. First, you need to familiarize yourself with the special signs, symbols, and terms that make up that language. The sections that follow offer an easy introduction to the "language" of algebra.

What Is Algebra?

Algebra is really nothing more than arithmetic—with one important exception: In arithmetic you use only numbers; in algebra you use both numbers and *letters*. A letter (x or y, or a or b, or any of the other 22 letters—whether capital or small, but customarily small) simply stands for a number that is unknown—a number whose value you have to figure out.

For example, in arithmetic you might see an equation like 3 + 5 = 8. But in algebra you might see 3 + x = 8. It's up to you find out which number x stands for. Now, that example is so simple that you see immediately that x must stand for 5.

But usually, in algebra, it's not immediately apparent which number a certain letter stands for (as in, say, 5.7 + x = 11.3). But even in a case like that, it's not difficult to *solve* for x (that is, to find the value of x) once you know how to do it.

Signs

The signs you use in arithmetic for adding, subtracting, multiplying, and dividing are also used in algebra. To review:

Addition:
To show addition, you use a *plus* sign: +
So, to show the addition of a and b, you write $a + b$.

Subtraction:
To show subtraction, you use a *minus* sign: −
Example: $a − b$.

Multiplication:
To show multiplication, you can use a *times* sign: x
Example: a x b.

To show the multiplication of two numerals (numbers), you always use the *times* sign (example: 5 x 4). But when multiplying a number by a letter or a letter by another letter, you actually have three different ways to show multiplication. First, you have the *times* sign: x (example: *a* x *b*).

Also, you can indicate multiplication with a dot (·) between the symbols (example: *a* · *b*).

Finally—and this is by far the most common way— you can show multiplication by simply placing a number and a letter, or a letter and another letter, right up against each other, with no space between them, as in 5*a* or *ab*.

So, 5 x *a*, 5 · *a*, and 5*a* all mean the same thing; they all mean 5 times *a*.

Note: The reason arithmetic doesn't use the second or third method mentioned above is that their use would cause confusion. For example, if you show the multiplication of 5 and 4 as 5 · 4, that could be mistaken (especially when handwritten) as 5.4 (five and four-tenths). And if you write the numbers right up against each other (54), it indicates the number fifty-four!

Division:
To show division, you can use a *divided by* sign ÷

Example: $a \div b$.

Also, you can use a fraction bar: —— , or (using the slanted version): /
Example: a/b.

So, $a \div b$ and a/b both mean the same thing; they both mean a divided by b.

Algebraic Expression

An *algebraic expression* is any collection of letters with signs between them. Thus, $a + b + c - d$ and $5ab + 2cd - 3ef$ are examples of algebraic expressions.

Terms

The *terms* of an algebraic expression are the different parts, separated by the signs. So, in the expression $5ab + 2cd - 3ef$, the individual terms are $5ab$, $2cd$, and $3ef$.

Monomial, Binomial, Trinomial, Polynomial

An algebraic expression of one term only (for example: $5ab$) is called a *monomial* (*mono* is the Greek prefix meaning *one*). So, a *term* and a *monomial* are really the same thing. An expression of two terms (for example: $5ab + 2cd$) is a *binomial* (*bi* is the Greek prefix meaning *two*). An expression of three terms (for example: $5ab + 2cd + 3ef$) is a *trinomial* (*tri* is the Greek prefix meaning *three*). But in general, any expression of more than one term can be called a *polynomial* (*poly* is the Greek prefix meaning *many*).

Factor

In arithmetic, you can multiply two smaller numbers together to produce one bigger number. For example: 5 x 7 = 35. Each of those smaller numbers, 5 and 7, is a *factor* of 35. Or, to put it another way, a *factor* is a number by which another larger number is exactly divisible (35 is exactly divisible by 5 and by 7, so 5 and 7 are factors of 35). In algebra, *factor* means the same thing as in arithmetic, but involves letters. So, if x x y = z, then x and y are factors of z. Likewise, a, b, and c are the factors of abc.

Exponent

Sometimes in a term, the same factor appears several times, as in a x a x a x a, or $aaaa$. For the sake of brevity, we use a shortcut to describe and write such a term. Instead of saying "a multiplied by itself four times," we say "a to the fourth *power*" (or simply "a to the fourth"). And instead of writing $aaaa$, we write a^4. The numeral above and to the right of a letter (or number), which shows how many times that letter (or number) is taken as a factor (in this case, the numeral 4 above and to the right of the a), is called an *exponent*.

Note: If a letter is taken only twice as a factor, as in aa, we write a^2 and we commonly call that "a squared" rather than "a to the second power" (though each phrase means the same thing). And if a letter is taken as a factor three times, as in aaa, we write a^3, and we commonly call that "*a* cubed" rather than "*a* to the third power."

If a letter has no exponent attached to it, it is *understood* that the exponent is 1, which means that that letter is taken as a factor only one time; in other words, there is no difference whatsoever between *a* and a^1 (each means *a*, one time).

Positive and Negative Quantities

Any term that has a *plus* sign, or no sign, before it is considered *positive* (that is, greater than zero). Any term with a *minus* sign before it is considered *negative* (less than zero). For example, +2 (or simply 2) is two greater than zero, and –2 is two less than zero. Likewise, using letters, *a* and +*a* are positive, while –*a* is negative.

Like and Unlike Signs

When two terms are both *plus* (positive), or both *minus* (negative), they are said to have *like* signs. When one term is *plus* and the other *minus*, they are said to have *unlike* signs.

Coefficient

The expression *a* + *a* + *a* + *a* means "*a*, four times," which is the same as *a* x 4, or 4*a*. In the term 4*a*, 4 is called the coefficient of *a*. Similarly, *ab* + *ab* + *ab* = 3*ab*, and 3 is the coefficient of the product *ab*. So, a *coefficient* is the number written before a letter, or letters, or a quantity, to show the number of times it is taken.

If a letter has no coefficient before it, it is *understood* that the coefficient is 1, which means that there is only one of that letter; in other words, there is no difference whatsoever between *a* and 1*a* (each means *a*, one time).

Like and Unlike Terms

Like terms are those which differ only in their coefficients. All others are *unlike*. So, 7*a*, 8*a*, and 5*a* are like terms; similarly, 8*a*^{2} and 6*a*^{2} are like terms (in each case, the coefficients are different, but the letters after the coefficients are the same). But 5*a* and 5*b* are unlike terms (the letters after the coefficients are different).

2 COMBINING LIKE TERMS

A group of like terms can be combined into a single term simply by adding their coefficients. For example, let's say you want to add together (that is, to simplify into a single term) the following terms: $2a$, $6a$, $4a$, and a (which, as an algebraic expression, is written $2a + 6a + 4a + a$).

First, remember that even though the final term, a, has no coefficient, its coefficient is understood to be 1. Adding the coefficients of the terms in our example gives you 13 ($2 + 6 + 4 + 1 = 13$). So the sum of the terms is $13a$. Note that the a's themselves are not added together to become $4a$. They stay as simply a in the answer because a's are the things you're determining the quantity of (and you have 13 of them). This concept is easy to grasp when you realize that if you hold two pennies in your right hand and three pennies in you left, you have five pennies.

Sometimes all the terms you want to combine are negative. If that is the case, simply add the coefficients and place a minus sign before the sum. For example, say you want to combine the following terms: *–4ab*, *–3ab*, *–2ab*, and *–ab* (which, as an algebraic expression, is written *–4ab – 3ab – 2ab – ab*). Adding the coefficients gives you 10 (4 + 3 + 2 + 1 = 10). Since all the coefficients are negative, place a minus sign before the 10; finally, write the common letters after the coefficient, thusly: *–10ab*.

Note: You'll remember from arithmetic that subtracting a quantity with *no* sign is the same as subtracting a quantity with a *plus* sign; that is, 3 – 2 is exactly the same as 3 – (+2). You'll also remember that *subtracting* a *positive* quantity is exactly the same as *adding* a *negative* quantity; that is, 3 – (+2) is exactly the same as 3 + (–2), each of which can be written as simply 3 – 2. So, as an algebraic expression, our example above could also be written *–4ab* + (*–3ab*) + (*–2ab*) + (*–ab*). This form of the expression is more cumbersome to write, but it shows more clearly that you are indeed *adding* together *negative* coefficients.

Sometimes you want to combine like terms, some of which are *positive* and some of which are *negative*. Let's say you want to combine the following: 5*xy*, –8*xy*, 3*xy*, and –2*xy*, which, as an algebraic expression, is written $5xy - 8xy + 3xy - 2xy$. Here are the steps to take:

Step 1: Add together only the positive coefficients. Remember that if a coefficient has no sign before it, it is *understood* to be positive (so $5xy$ has a positive coefficient). The positive coefficients add up to 8 ($5 + 3 = 8$).

Step 2: Add together only the negative coefficients. The negative coefficients total 10 ($8 + 2 = 10$).

Step 3: Find the difference between the two sums. This gives you a single, combined coefficient (but without a sign attached to it yet). The difference between 10 and 8 is 2 ($10 - 8 = 2$), giving you a combined coefficient of 2.

Step 4: Before the coefficient, place the sign (*plus* or *minus*) of the terms whose sum is greater. In this example, the *negative* terms have the greater sum (10 is greater than 8). So the coefficient takes a *minus* sign: -2. (Note: If the positive terms have a greater sum, you don't have to write a plus sign before the coefficient because a number with no sign is understood to be positive.)

Step 5: Attach the common letter or letters after the coefficient. The common letters are xy, so the answer is $-2xy$.

Practice problems
In each examples below, combine the like terms into a single term.

1. $4x + x + 9x + 11x =$

2. $-5ax - 4ax - ax - 2ax =$

3. $10a^2 + 3a^2 - 4a^2 =$

4. $12cx^2 - 4cx^2 - 9cx^2 =$

Note that *unlike* terms cannot be combined into a single term. Imagine that you have two nickels in your right hand and three dimes in your left. You don't have five of any particular type of coin. Likewise, if you want to combine $2x$ and $3y$, you don't have five of any particular letter. All you can do to combine the unlike terms is to create a polynomial with the proper sign, thusly: $2x + 3y$.

However, if an expression contains both like and unlike terms, you can simplify the expression by combining just the like terms. Thus, $5a + 2b - 3a + 4b$ can be written more simply as $2a + 6b$ (because $5a$ and $-3a$ equal $2a$, and $2b$ and $4b$ equal $6b$).

In the examples below, reduce each polynomial to its simplest form by combining like terms. Note: Numbers with no letters attached to them are combined as in arithmetic (for example, $+10$ and -3 can be written simply as 7).

5. $5a - 3b + 10 + 6a - 8b - 3 =$

6. $6a - 2b + 5c + 4a + 8b - 2c =$

7. $6x^2 - 8x + 1 + 3x^2 - x + 6 =$

8. $-21a^2 - 14ab + 20ac^2 + 45a^2 - 20ab - 12ac^2 =$

3 MULTIPLYING TERMS

Multiplying a Term by a Number

We know that $3a + 3a$ is $3a$ two times. In other words, adding a quantity to itself a certain number of times is the same thing as multiplying that quantity by that number of times. So, $3a + 3a$ is the same as $3a$ x 2 (each equals $6a$). As a rule, to multiply a term by a number, multiply the term's coefficient by that number and keep the term's letter or letters as they are. So, for example, 4 x $3xy = 12xy$ (just as $3xy + 3xy$ $+3xy +3xy = 12xy$).

Multiplying Unlike Terms

If you see an expression like $3a$ x $2b$, you may wonder how something can be multiplied by something other than a simple number. You know what it means to take (multiply) something two times (as when, in the example above, we took $3a$ two times). But what does it mean to take something $2b$ times? If you realize that the letter b actually stands for a number whose value is not known, then you realize that $2b$ also represents a number. So, if, for example, b happens to represent the number 5, then $2b$ represents the number 10, and we'd be talking about taking $3a$ ten times.

But if we don't know the value of b, then b must stay as a letter (but we understand that multiplying a quantity by $2b$ is meaningful). As a rule, to multiply one term by another, multiply the coefficients by each other and multiply the letters by each other. Thus, $3a$ x $2b = 6ab$.

Multiplying Powers of the Same Letter

We know that a^4 is the same as $aaaa$, and a^3 is the same as aaa. So, multiplying a^4 by a^3 is the same as $aaaa$ x aaa, which, by the rule of placing letters to be multiplied together right up against each other with no sign between them, is $aaaaaaa$, which is the same as a to the seventh power, or a^7 It is easily seen that the exponent of the answer, 7, is the sum of the exponents of the terms $(4 + 3)$. As a rule, to multiply powers of the same letter, simply add the exponents. Of course, if you want to multiply together *different* letters that have exponents, such as a^2 x b^3, it is impossible to simplify or combine the terms, except to write the expression without the times sign, thusly: $a^2 b^3$.

Multiplying Terms with Unlike Signs

Sometimes the terms you want to multiply have unlike signs, as in $3a$ x $-2b$. As you remember from arithmetic, multiplying + by + gives +; multiplying – by – gives +; multiplying + by – gives –; and multiplying – by + gives –. Or, to put it another way, multiplying *like* signs gives + and multiplying *unlike* signs gives –. So,

$+3a$ x $+2b = +6ab$
$-3a$ x $-2b = +6ab$
$+3a$ x $-2b = -6ab$
$-3a$ x $+2b = -6ab$

To review the rules for multiplying one term by another:

For Coefficients: Use common multiplication.
For Letters: For like terms, keep the letter as is; for unlike terms, multiply the letters.
For Signs: Like signs make +; unlike signs make –.
For Exponents: Add the exponents of the same letters.

Examples:

$3a^2$ x $4a^3 = 12a^5$
Multiplying the coefficients, 3 and 4, gives 12. Multiplying like signs (both +) gives a product that is +. Adding the exponents, 2 and 3, gives 5.

$5ab \times 4ab^2c = 20a^2b^3c$

When no exponent is shown, the exponent is understood to be 1 (because a letter without an exponent is understood not to be multiplied by itself any number of times); thus, the terms to be multiplied are understood to be $5a^1b^1$ and $4a^1b^2c^1$. And by the rule of adding exponents of the same letters, the product contains powers (exponents) of 2 for a ($1 + 1 = 2$), and 3 for b ($1 + 2 = 3$).

$4a^4 \times -axy = -4a^5xy$

The coefficient of the second term, $-axy$, is understood to be 1, and 4 x 1 is 4. And because the terms have unlike signs, the sign of the product is –.

$-5x^2y^2z^2 \times -2x^2zy^3 = 10x^4y^3z^5$

It is easily seen that the coefficients have been multiplied and exponents (of like letters) have been added. Even though both terms being multiplied are negative, the product is positive (remember: like signs gives +).

$2ab \times -3cy \times -a^3b^2y = 6a^4b^3cy^2$

Here you have three terms being multiplied together rather than just two, but the process for determining the product is the same. Multiplying the coefficients gives 6 (2 x 3 x 1 = 6). (Remember: A term with no coefficient is understood to have a coefficient of 1.) To determine the sign of the product, note the signs of the individual terms. In this case you have +, –, and –, respectively. Multiplying the first two terms, which have unlike signs (+ for the first and – for the second), gives a partial product whose sign is – (because unlike signs make –). Then, multiplying together that partial product and the third term, which have like signs (– for the partial product and – for the third term), gives + (because like signs make +).

Practice problems
Find the product of each example below.

1. axy x bx =

2. $3ab$ x $-ax$ =

3. $-3mn$ x am =

4. $-xy^2$ x $-xy^2$ x $-11x$ =

4 MULTIPLYING EXPRESSIONS

Multiplying a Polynomial by a Monomial

Sometimes you have to multiply a *polynomial* (an expression of two or more terms) by a *monomial* (a single term). For example, let's say you have to multiply the polynomial $12a - 7b$ by the monomial $9a$. If the example is written on a single line, to show the polynomial as separate and distinguished from the monomial, we place parentheses around the polynomial, thusly: $(2a - 7b)$ x $9a$. As a rule, to multiply a polynomial by a monomial, multiply each term of the polynomial by the monomial, and combine (add together) the partial products. In our example, the partial products are $18a^2$ (because $2a$ x $9a = 18a^2$) and $-63ab$ (because $-7b$ x $9a = -63ab$). Combining the partial products gives $18a^2 + (-63ab)$, or, more simply, $18a^2 - 63ab$, which is the answer.

To multiply a three-term polynomial by a monomial, follow the same procedure; that is, multiply each term of the polynomial by the monomial. For example: $(2a - 3b +$

5c) x $-2abc = -4a^2bc + 6ab^2c - 10abc^2$. As a rule, no matter how many terms are included in the polynomial, multiply each by the monomial (and add together all the partial products). And, whenever possible, reduce the answer to its simplest form by combining like terms.

Practice problems
Find the product of each example below.

1. $(5b - 8a)$ x $-12a =$

2. $(ac + 2bc)$ x $3a =$

3. $(10x^2 - 5ax - 3a^2)$ x $4x^2 =$

4. $(2ax + 5by - 3cz + 4x^2)$ x $-7x^2 =$

Multiplying a Polynomial by a Polynomial
Sometimes you have to multiply a polynomial by another polynomial. As a rule, multiply *each* term of the first polynomial by *each* term of the second (and add together all the partial products). Then, if possible, reduce the answer to its simplest terms by combining like terms. For example, let's say you want to multiply $3a + 2b$ by $5a - 4b$. To multiply each term of the first polynomial by each term of the second polynomial means (in this example, which has only two terms in each polynomial) that you (1) multiply the *first* term of the first polynomial by the *first* term of the second polynomial; (2) multiply the *first* term of the first polynomial by the *second* term of the second polynomial; (3) multiply the *second* term of the first polynomial by the *first* term of the second polynomial; (4)

multiply the *second* term of the first polynomial by the *second* term of the second polynomial. This process gives us:

$(3a \times 5a) + (3a \times -4b) + (2b \times 5a) + (2b \times 4b)$, which is the same as:

$15a^2 + (-12ab) + 10ab + 8b^2$, which, combining like terms (the ab terms), is the same as:

$15a^2 - 2ab + 8b^2$, which is the answer.

No matter how many terms the polynomials that are being multiplied together contain, each term of the first polynomial must be multiplied by each term of the second. For example, multiplying $a + b + c$ by $x + y + z$ gives:

$ax + ay + az + bx + by + bz + cx + cy + cz$.

Practice problems
Find the product of each example below.

5. $(x + 3) \times (x - 2) =$

6. $(a^2 + 2x^2) \times (3a^2 + x^2) =$

7. $(x^3 + x^2 + x) \times (x - 1) =$

8. $(2a + 3b - 5c) \times (a + b - c) =$

5 DIVISION

Division is like multiplication in reverse. For example, in multiplication you say that 3 x 5 = 15, but in division you say that 15 ÷ 5 = 3 (or that 15 ÷ 3 = 5). Or, to put it into words, to *multiply* is to obtain, from a number, another number that contains that number a specified number of times (as in our example, where we obtained 15 by taking the quantity 3 exactly five times). To *divide* is to find out how many times a certain number contains another number (as in our example, where we saw that the number 15 contains the number 5 three times, or that it contains the number 3 five times).

Using letters now instead of numbers, we can understand that:

a x b = ab; hence $ab ÷ a = b$ and $ab ÷ b = a$.

We also know that that:

$a \times (-b) = -ab$; hence $-ab \div a = -b$ and $-ab \div (-b) = a$.

$-a \times b = -ab$; hence $-ab \div (-a) = b$ and $-ab \div b = -a$.

$-a \times (-b) = ab$; hence $ab \div -a = -b$ and $ab \div -b = -a$.

From these examples, we can derive a rule for signs in division; namely (just as in multiplication), like signs give + and unlike signs give −.

Concerning exponents in division, we know that:

$a^3 \times a^4 = a^7$; hence, $a^7 \div a^4 = a^3$ and $a^7 \div a^3 = a^4$.

Therefore, we see that to divide two powers of the same letter, we *subtract* exponents (as expected, this is just the opposite of what we do when multiplying powers of the same letter; namely, *add* exponents).

We can now state the rules for dividing one term by another:

For Coefficients: Use common division.
For Letters: For like terms, keep the letter(s) as is; for unlike terms, divide the letters.
For Signs: Like signs make +; unlike signs make −.
For Exponents: Subtract the exponents of the same letters.

Note: In an example such as $a^2 \div a^2$, you know right away that the answer is 1, because (as you remember from arithmetic) any quantity divided by itself is 1. But by the rule for division of exponents stated above, your answer is a^0 (*a* to the zero power), because, subtracting exponents, 2 − 2 = 0. What is important to realize is that there is no contradiction between the two answers. Why? Because—and this is a rule of mathematics that must be memorized—*any quantity taken to the zero power equals 1.*

This can be proven—or at least easily demonstrated—if you consider the various "columns" used in writing regular numbers; that is, the *ones* column, the *tens* column, the *hundreds* column, the *thousands* column, and so on. As you remember from arithmetic, if you write a number like, say, 5,324, you have five thousand (because 5 is in the thousands column), three hundred (because 3 is in the hundreds column), and twenty-four (because 2 is in the tens column and 4 is in the ones column).

Now let's take the numbers that make up those columns (1000, 100, 10, and 1) and rewrite them as powers of 10:

1000 can be written as 10 x 10 x 10, or 10 to the 3rd, or 10^3.
100 can be written as 10 x 10, or 10 to the 2nd, or 10^2.

10 can be written as 10 to the 1st, or 10^1.
1 can be written as 10 to the zero, or 10^0.

You see that as we move down the list, each number is one power of 10 less than the one above; hence, the exponents, as we move down the list, drop from 3 to 2 to 1, and so on. The last number in the list, 1, representing the ones column, if written as a power of 10, must be 10^0 in order to follow the established pattern of descending exponents.

Lest you think this demonstration—of a number to the zero power equaling 1—works only for powers of 10, let's try some other number. In the list below, in which each successive row is one power less of 6 (a number arbitrarily chosen for demonstration), note that in order to preserve the established pattern of descending exponents, 1 must be written as 6^0.

216 can be written as 6 x 6 x 6, or 6 to the 3rd, or 6^3.
36 can be written as 6 x 6, or 6 to the 2nd, or 6^2.
6 can be written as 6 to the 1st, or 6^1.
1 can be written as 6 to the zero, or 6^0.

No matter what number is chosen for demonstration, as the powers of that number descend finally to 1, the exponent of that final number—to preserve the pattern of descending exponents—must always be zero. So, to state the rule once more, *any quantity taken to the zero power equals 1*. And that's why when you see

$a^2 \div a^2$, you know that the answer is 1 *either* because of the rule that anything divided by itself is 1 *or* because of the rule that when you divide one term by another, you subtract exponents (here, $2 - 2 = 0$, giving you a^0, which equals 1).

Also note: A "trick" you may be aware of is that when you have a fraction such as ab/a, because a is a factor of both the top and bottom of the fraction, you can "cancel out" the a's, giving you an answer of simply b. Why can you do this? Actually, there are two reasons. First, you know that any quantity divided by itself is 1. Hence, the fraction is the same as 1 x b (because $a \div a = 1$), and thus the fraction can be written as $1b$, or simply b (because a letter with a coefficient of 1 is the same as that letter with no coefficient). Second, you know from the rule of subtracting exponents when dividing terms (and, after all, a fraction, by definition, is really a division problem, in with the top quantity is divided by the bottom quantity) that our fraction can be written as a^0 x b (because subtracting the exponents of a gives you $1 - 1 = 0$). And because a^0 is equal to 1 (as is *any* quantity to the zero power), we can now change a^0 x b to 1 x b, or $1b$, or simply b.

So, in the list of rules stated earlier for dividing one term by another, concerning coefficients and exponents, it would be more accurate to state:

For Coefficients: Use common division (or simply "cancel out" coefficients that are the same). So, for example, by canceling out the 5's, $5a/5b$ becomes a/b.
For Exponents: Subtract the exponents of the same letters (or simply "cancel out" letters with the same exponents). So, for example, by canceling out the a^2's, a^2b/a^2 becomes b.

Examples:

$12xy \div 3x = 4y$
Dividing coefficients gives 4. Because the *dividend* (the technical term for the quantity being divided in a division problem, or the term for the top portion of a fraction) and the *divisor* (the technical term for the number the dividend is divided by, or the term for the bottom portion of a fraction) both contain x, the x's can be cancelled out, leaving y.

$abc \div -a = -bc$
Because each term has no coefficient (or has the unwritten but understood coefficient of 1), the coefficients can be ignored (or canceled out); hence, the *quotient* (the technical term for the answer to a division problem) has no coefficient (or it has the unwritten but understood coefficient of 1). Canceling out the a's leaves bc. Because the signs of the dividend and divisor are unlike, the sign of the quotient is $-$.

$-15ab^2c \div -3ab = 5bc$

Dividing coefficients gives 5. Cancel the a's. For the b's, subtract exponents (2 – 1 = 1), leaving b^1, or simply b. Like signs make + (which is understood but not written in the quotient).

$-4a^4b^3c^5 \div -a^2bc^2 = 4a^2b^2c^3$

Dividing coefficients gives 4 (4 ÷ 1 = 4). The sign of the quotients is + because like signs make +. For the letters and their exponents, subtract exponents: for a, 4 – 2 = 2; for b, 3 – 1 = 2; for c, 5 – 2 = 3.

Practice problems

Find the quotient of each example below.

$-axy \div y =$

$-8axy \div 8ax =$

$-20a^2b^3c^4 \div 5abc =$

$14a^3xy \div 7ay =$

Sometimes a polynomial (an expression of more than one term) must be divided by a monomial (a single term). In such a case, simply divide the monomial into each term of the polynomial.

Examples:

$(40x^2 - 24ax) \div 8x = 5x - 3a$
Dividing $40x^2$ by $8x$ gives $5x$, and dividing $-24ax$ by $8x$ gives $-3a$.

$(12x^3 - 21ax^2 + 3a^2x) \div 3x = 4x^2 - 7ax + a^2$
Dividing $12x^3$ by $3x$ gives $4x^2$; dividing $-21ax^2$ by $3x$ gives $-7ax$; and dividing $3a^2x$ by $3x$ gives a^2.

Practice problems

$(2ab + 3ac - 4ad) \div a =$

$(ax + bx^2 - cxy) \div -x =$

$(6a^2x^2 - 8abx + 2ax^2) \div -2ax =$

$(5bc + 35abc^2 - 10b^2c^2) \div 5bc =$

6 FACTORING

A *factor* is a quantity by which another quantity is exactly divisible. And, as a verb, to *factor* a quantity or expression is to break it into pieces, which, when multiplied together, will produce that original quantity or expression (as a product). For example, the term 5*a* can be factored into two separate terms: 5 and *a* (because 5 x *a* gives the product 5*a*). And 15*ab* can be factored into 3, 5, *a*, and *b* (because 3 x 5 x *a* x *b* = 15*ab*).

Dividing a Polynomial into Factors, One of Which Is a Monomial

Often, simply by looking at a polynomial and by understanding the rules of multiplication and division, you can factor the polynomial into one monomial and one simpler polynomial in your head. As a rule, to do so, simply divide the given polynomial by the greatest monomial that will exactly divide each of its terms.

For example, let's say you want to factor the quantity 2*ax*

+ $2ay + 2az$ into a one monomial and one simpler polynomial. You see at once that each term of the given polynomial is exactly divisible by $2a$. (Note: It's true that each term of the polynomial is also exactly divisible by 2 and by a; but $2a$ is the *greatest* monomial that will exactly divide each term.) So, one factor of the given quantity, the monomial factor, is $2a$. The other factor, the simpler polynomial factor, is what is left of the original quantity after each of its terms has been divided by $2a$; namely, $x + y + z$. So, the factors of $2ax + 2ay + 2az$ are $2a$ and $(x + y + z)$. Or, to state it mathematically, $2ax + 2ay + 2az = 2a (x + y + z)$.

To check that you've factored a quantity correctly, simply multiply the factors together. If the product is the same as the original quantity, you've factored correctly. Using our example, $2a (x + y + z)$, you see that if you multiply the factors, you get $2ax + 2ay + 2az$; so our factoring is correct.

More examples of dividing a polynomial into factors, one of which is a monomial:

$bc^2 + bcd = bc (c + d)$
By inspection, you see that the greatest common factor is bc (that is, each term of the given quantity is exactly divisible by bc). Dividing bc into each of the terms of the given quantity gives $c + d$. So, the factors are bc and $(c + d)$. Multiply the factors together, and you see that your product is the original quantity.

$6ax^2y + 9bxy^2 - 12cx^2y = 3xy(2ax + 3by - 4cx)$

The greatest common factor is $3xy$. Dividing that into each term of the given quantity gives $2ax + 3by - 4cx$.

Practice problems

Factor each expression into one monomial and one polynomial:

$4x^2 + 6xy =$

$5ax^2 - 35ax^3y + 5a^2x^3y =$

$a^3cm^2 + a^2c^2m^2 - a^2cm^3 =$

Factoring a Quadratic Trinomial into two Binomials

A *quadratic* expression is one that involves an unknown quantity (represented by a letter, such as x, for example) taken to the second power (but to no higher power than the second). And, of course, a *trinomial* is an expression made up of three terms. So, a typical *quadratic trinomial* might look like this: $x^2 + x + 1$, or $3x^2 + 5x - 7$, or $12x^2 - 6x - 5$. You see that in each case, the first term is x^2 multiplied by some coefficient (which may or may not be the coefficient 1); the second term is simply x multiplied by some coefficient (which may or may not be the coefficient 1); and the third term is simply a number (with no x appearing).

Note: You can refer to each of the members of the quadratic trinomial in terms of their exponents (of x). The first term always contains x^2; the second always contains

x^1, which is the same as (and is written as) x; and the third term always contains x^0, which is the same as 1 (and is simply understood and thus not written). So, you can think of each of the actual *numbers* of a quadratic trinomial (even the final number) as *coefficients*. For instance, in our last example above, $12x^2 - 6x - 5$, if you think of that as $12x^2 - 6x^1 - 5x^0$, you see that the numbers—12, 6, and 5—are indeed coefficients.

Now, by substituting the letters a, b, and c for the respective coefficients, we can state a general rule: The quadratic trinomial is of the form $ax^2 + bx + c$, in which x is a *variable* (an unknown quantity) and a, b, and c represent *constants* (actual numbers).

Note: If a coefficient happens to be zero, its term disappears (because anything multiplied by zero is zero). Hence, it is possible to have a quadratic *binomial* (as when the coefficient of the second or third term is zero), such as $x^2 - x$ or $x^2 + 2$. Also note: If the coefficient of the first term is zero, you are left with a binomial, such as $3x + 5$, which is no longer quadratic (because there is no variable taken to the second power). Such an expression (in which the variable has no exponent) is called a *linear* expression (so, our example, $3x + 5$, is a *linear binomial*).

Unlike factoring a polynomial into a monomial and a simpler polynomial, there is no general rule for easily factoring a quadratic trinomial into two binomials. Doing so takes a certain amount of logic, intuition, and trial and error. But coming up with the right answer can be rewarding—and fun. Although these types of problems are beyond the scope of this simple primer, you can find them, along with hints for solving them, in any standard algebra textbook.

www.ingramcontent.com/pod-product-compliance
Lightning Source LLC
Chambersburg PA
CBHW070948210326
41520CB00021B/7112